Issaka Mahaman Dalibou

Usefulness of endangered plant species

Issaka Mahaman Dalibou

Usefulness of endangered plant species

in rural Niger: the case of the baobab " Adansonia digitata " in Kotaki and Banigoungou

ScienciaScripts

Imprint

Cover image: www.ingimage.com

This book is a translation from the original published under ISBN 978-620-6-73466-6.

Publisher:
Sciencia Scripts
is a trademark of
Dodo Books Indian Ocean Ltd. and OmniScriptum S.R.L publishing group

120 High Road, East Finchley, London, N2 9ED, United Kingdom
Str. Armeneasca 28/1, office 1, Chisinau MD-2012, Republic of Moldova, Europe
Managing Directors: Ieva Konstantinova, Victoria Ursu
info@omniscriptum.com

Printed at: see last page
ISBN: 978-620-8-98970-5

The usefulness of endangered plant species in rural areas of Niger: the case of the baobab "*Adansonia digitata*" in Niger. Kotaki and Banigoungou

ISSAKA MAHAMAN Dalibou
Abdou Moumouni University, Niamey (Niger).
Geography Department
Laboratoire d'Études et de Recherches sur les Territoires Sahélo-Sahariens (LERTESS).

TABLE OF CONTENTS

Summary

The baobab is one of the main woody food species of Sahelian populations. Today, this species is threatened with extinction in several countries as a result of destructive human and natural actions. The aim of this study is to show the impact of man's actions, on the one hand, and nature's, on the other, on the baobab. Two (02) localities were used as testing grounds to operationalize the present study. These were Kotaki and Banigoungou. The method used was empirical. A questionnaire was used to interview 31 baobab-holders aged 49 and over. A semi-directive interview guide was also used with 03 technical managers and 03 representatives of the local associations involved. The data collected highlight not only the virtues but also the threats faced by the species studied in Kotaki and Banigoungou. Key words: Adansonia Digitata, Banigoungou, climate change, Kotaki

Abstract

The baobab is one of the main woody food species of Sahelian populations. This species is today threatened with extinction in several countries following destructive anthropogenic and natural actions. This study aims to show the weight of human actions on the one hand, but also that of nature, on the other, on the baobab. Two (02) localities served as experimental grounds to make this study operational. These are Kotaki and Banigoungou. When it comes to field work, the method used is empirical. The field survey via a questionnaire made it possible to interview 31 people with baobabs whose ages ranged from 49 and over. Also, a semi-directive interview guide was passed to 3 technical executives concerned and 3 representatives of the local associations involved. The data collected highlights not only the virtues but also the threats faced by the species studied in Kotaki and Banigoungou.

Introduction

The baobab is native to tropical Africa. It is mentioned as early as 1354 in Ibn BATTUTA's accounts of his travels. The origin of the word "Baobab" can be found in the work of ALPINO, A. GABAR DIOP et al. (2005, p.2).

Indeed, the word baobab is a transliteration of the Arabic word "bu hibab", meaning "fruit with many seeds". Other works refer to the nutritional value of the species in question on its content (A. MUSA 2024).

We learn from the FAO (1995) that baobab bark contains excellent quality fiber and produces vitamin-rich fruit.

In addition to its nutritional value, the leaves and fruits of this species are traded by farmers, which is why people in the areas studied are more interested in preserving this plant variety as much as possible. Despite these advantages, the species is under real threat, compromising its development and increasing the risk of extinction.

Indeed, the study's experimental areas reveal that the baobab is now threatened with extinction as a result of human deforestation and/or clearing.

Socio-environmental and ecological constraints linked to climate change, such as recurrent droughts, have a negative impact on the future of this tree species.

So, the fundamental question posed by the present reflection is the following: Why is the baobab in a state of extinction in Kotaki and Banigoungou?

To answer this main question, a hypothesis was formulated according to which the baobab faces abusive cutting of branches, bark and roots, followed by effective collection of fruit and leaves.

To this end, this work focuses on the characteristics and virtues of the baobab and the real actions leading to its extinction in the two (02) villages studied.

1. Methodology

The methodology used in this study is based on a combination of qualitative and quantitative methods. Data collection took place simultaneously in Kotaki and Banigoungou from March 7 to 21, 2024.

The study also relied on cartography using ArcGIS software (version 10.8) to better illustrate the localities involved, with a view to enhancing the value of the data collected.

1.1. Location of study area

The localities of Kotaki and Banigoungou are administratively attached to the Falmey department. They are bordered on either side by the Fabridjie and Kankandi rural districts, on the east by the Golley and Sambera rural districts, on the west by the Parc de W and the Kirtachi rural district, and on the south by the Republic of Benin.

Map 1: Geographical location of the study area

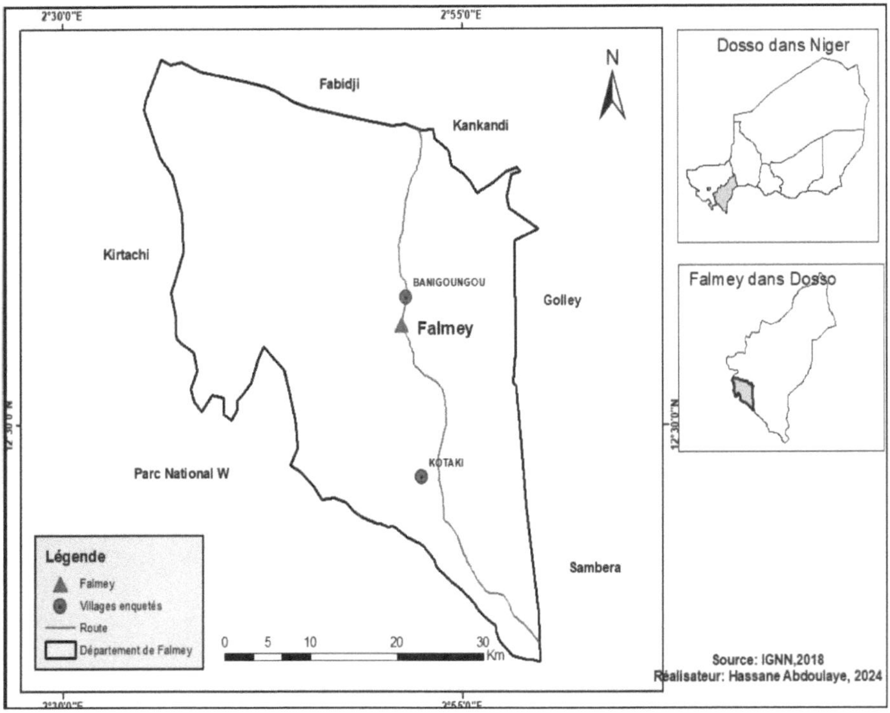

1.2. Theoretical and methodological considerations

The social sciences and humanities are characterized by their reliance on theory, and their ability to cross-reference data with empirical evidence. This is as true for disciplines such as Human Geography as it is for Sociology or Social and Cultural Anthropology.

The choice of actors depends on the particularities of the area under investigation, the subject under study and the key informants.

According to S. Paugam, (2010), generally, the researcher asks himself three preliminary questions to carry out an investigation in the field of social sciences:

1. Who should be surveyed?
2. Where?
3. And what's the approach?

At this level, we have exploited the documentation on the disappearance and use of baobab extracts, as well as field work.

1.3. **Literature search**

The grey literature mobilized to build the theoretical dimension relates to the documentation that takes up the questions of the risk of disappearance of the species; and the need to protect the tree species.

As a general rule, the documentary search examined themes directly or indirectly linked to the object of study in question, in order to determine the overall situation. This work was carried out by consulting documents in the library of the Geography Department of the Abdou Moumouni University in Niamey, the DOUDDAL digital library, scientific articles, NGO and project reports and websites such as futura-sciences.com.

1.4. **Comments**

As part of this work, direct observations were carried out in the field to better discern the extent of the disappearance of the baobab plant in the rural commune of Falmey. Thus, the primary objective of the field visit was to observe the effects of anthropic actions on the baobab and the activities actually carried out by a few actors for the benefit of the inhabitants.

1.5. **Maintenance**

As Alain Blanchet and Anne Gotman (2007, p. 24) point out, the interview is the most suitable method for gathering the "subjectively intended meaning" (Weber, 1998) or "reasons" (Boudon, 2003) of respondents:

Also known as "guided interviews" or "semi-structured interviews", semi-structured interviews are widely used in the humanities and social sciences. "The semi-directive interview combines a non-directive attitude to encourage the exploration of thought in a climate of trust and a directive project to obtain information on points defined in advance" (Berthier, 2010, p. 78).

Semi-directive interviewing is more flexible than directive interviewing, since the interview guide is just as structured, and the interviewer will ask questions in the order best suited to the respondent's discourse (Quivy and Van Campenhoudt, 2006).

The interviewer can also opt for a thematic interview guide with a list of themes to be addressed, without pre-written questions (Desanti and Cardon, 2010). This type of interview is appropriate when you want to explore a specific, circumscribed area in greater depth, explore hypotheses that are not all definitive, and invite the respondent to express him/herself freely within a framework defined by the interviewer.

A semi-structured interview guide of the Knowledge Attitude and Practice (CAP) type was designed to collect qualitative data from 3 managers involved in baobab protection and regeneration and 3 representatives of local associations.

The aim of this interview guide is to assess the state of disappearance of the baobab in Kotaki and Banigoungou.

These are :

- Direction Départementale de l'Environnement et de la Lutte Contre la Désertification (DDE/LCD),
- Direction Départementale de l'Agriculture (DDA),
- The Direction Départementale de l'Elevage (DDEl) of the Falmey department.

1.6. Questionnaire survey

In order to collect quantitative data for this study, a questionnaire was sent to baobab owners in the selected villages. Data was collected from people aged 49 and over.

With this in mind, we used the snowball technique, with the following inclusion criteria: being the owner of the plant, being 49 years of age or older, and the repetition of data justifying its extinction.

This empirical method consists of identifying one owner who may lead to another and so on until the data on plant extinction is repeated.

During fieldwork, 31 landowners were interviewed, including 11 in Banigoungou and 20 in Kotaki.

1.7. **Field data processing and analysis**

Before being analyzed, the data were first sorted and separated by locality, then checked for completeness and consistency. Checks were carried out mainly in the field during data collection, with the aim of detecting any errors made when filling in the questionnaire.

Each piece of data was then coded and recorded on a summary sheet.

Analysis was carried out using an Excel spreadsheet to produce graphical illustrations. As for the qualitative data, the interviews were recorded in local languages using digital dictaphones, and then translated into French.

The translated data were transcribed onto Word software, along with an analysis of the ambience of the unfolding and highlights of the interviews. The data was processed in the form of a thematic content analysis of the discourse delivered by each actor.

2. **Results presentation**

The data collected deals with the endangered species in the Sudanian zone, more specifically in Kotaki and Banigoungou. The aim is to present the characteristics, uses, extinction status and impacts of the baobab's disappearance.

2.1. **The story of the African baobab**

The history of the African baobab is well documented by Baum (1995). Linnaeus gave the binomial *Adansonia digitata* the generic name to honor Michel Adanson, who was the first botanist to describe the African baobab in the eighteenth century (M. Adanson, 1771).

The species belongs to the Malvaceae botanical family. The African baobab (*Adansonia digitata*) is a large tree, 18 to 25 m high, with a diameter of 5 m to 10 m.

- The trunk is irregular and shaped like a pot-bellied bottle.
- The crown is spreading, with huge branches and leaves clustered at the ends of the branches.
- The bark is smooth, gray with blue or purplish highlights.
- The baobab produces hanging flowers and capsules containing numerous seeds surrounded by a floury pulp.

2.2. **Main characteristics of baobab species**

The Adansonia genus belongs to the Bombacaceae family and the Malvaceae order, in the same way as the kapok tree and the cheese tree. It comprises eight (8) species, six (6) of which are endemic to Madagascar (A. ASSOGBADJO, et al, 2011, p.6). The same type of species can be found in northwestern Australia. Adansonia is a tree species with a crown, usually compact, 5 to 30 m high, and a trunk 2 to 10 m in diameter.

2.3. **The nutritional and medicinal virtues of baobab**

The baobab is a tree with multiple therapeutic uses. Every part of the tree (roots, pulp, bark, leaves, flowers, seeds, etc.), alone or in association with other plant species, is of definite use to people in many fields.

In pharmacopoeia, the organs and parts of the baobab are used to treat at least 19 different illnesses and conditions. The pulp is by far the organ of the baobab that contributes most to medicinal use, and is involved in the treatment of several ailments, the most important of which are anemia, malaria and diarrhea. The bark

helps to heal wounds, while the kernel extracted from the seeds is used to calm hiccups (nausea).
The leaves are used to treat hemorrhoids, while the root is used to treat epilepsy. All these recognized uses of the different organs of the species are obviously due to their richness in nutritive components and active substances.
As a result, alkaloids, flavonoids, sterols, coumarins and saponoids have been identified in the various organs of this species. These are nitrogenous, basic organic substances with physiological properties that support the nervous system and spinal cord (alkaloids). They are also venous tonics with antispasmodic, antiulcer and anti-inflammatory properties (flavonoids).
As polyphenolics, they combine with skin proteins to make the leather rot-proof (tannins). The presence of adansonine (C48H36O33) in the bark justifies its use against malaria and other fevers (Sidibé & Williams, 2002).
The very high iron content in both leaves and seeds, around 29.3% of dry matter, justifies its use in the treatment of anemia, as iron has the property of binding hemoglobin. The baobab population is overflowing with extremely important properties. As a result, these properties are not only food and nutritional, but also pharmaceutical. Whether it's its leaves, pulp, roots, seeds, flower or bark, the baobab represents a rich and calorific source of nourishment for mankind.

2.3.1 The properties of pulp

The pulp is the most valuable part. It is acidic, rich in ascorbic acid and contains high levels of calcium and magnesium. As the biochemical characterization of this product (pulp) is still incomplete, particularly in terms of polyosides, organic acids, aroma compounds or phenolic compounds, the pulp is often used as an antidote for fatigue, inappetence, lack of energy or lack of energy. They are used as an antidote for fatigue, inappetence, diarrhea, enteralgia, malaria, nasopharyngeal infections, circulatory disorders and insect bites. The white pulp extracted from baobab fruit is used in a range of hot and cold beverages. What's more, these extracts are very rich in vitamin C, estimated at more than 10 times the content of oranges, while the seeds also contain oil that is consumed by humans and rich in other elements.

Photography 1 : Baobab pulp

Source: Google image, 2024

> *"(...), I have four feet of baobab just for the consumption of its pulp. Beyond that, I sell the surplus to meet my needs"*. (Extract from an interview with a farmer in Banigoungou, March 10, 2024).

2.3.2 The properties of fruit

They are an antidote for fatigue, inappetence, diarrhea, enteralgia, malaria, nasopharyngeal infections, circulatory disorders and insect bites.

The testimony of a farmer confirms this assertion: *"the fruit of the baobab has several virtues, including fresh or hot drinks, etc."*. (Interview conducted on March 8, 2024).

Photo 2 : baobab fruit

Source: SEBASTIEN Garnaud, June 2006

2.3.3 Seed properties

The seeds are useful as a protein supplement. They are anti-diarrheal, anti-malarial and cure coughs. As baobab seeds are very rich in lipids (28%), protein (34%), minerals and vitamins, their by-products are potential sources of nutrients that can contribute to the nutritional security of populations, especially children in high-risk environments.

Photo 3 : Adansonia digitata seeds

Source : www.ethnoplants.com

2.3.4 The properties of leaves

Baobab leaves are used fresh or dried in gastronomy. They are rich in micronutrients, including vitamin A, calcium, iron, potassium, magnesium, manganese and phosphorus.

They also have therapeutic virtues and are used in the treatment of antihistamines, analgesics, disinfectants and anti-inflammatories, asthma, hypertension, circulatory disorders (hemorrhoids), enteralgia, aphrodisiacs, infant dental pain, perspiration activators, rheumatism, conjunctivitis, urinary infections, insect bites, dracunculiasis, skin inflammation.

> *"I have a field right next to the village. I planted two baobabs 10 years ago. It took me a long time to get the trees to start leafing out. Thank God, my investment was not in vain. It's the leaves that interest me, because they're gold on the local market. They are used to cook baobab leaf sauce, reputed to be rich in iron and antioxidants. Admittedly, the harvest is seasonal, but the added value makes it easy for me to reap the financial rewards. The same leaves are used to make decoctions to treat a number of illnesses. For example, they cure jaundice detected at an early stage".* (Extract from an interview conducted on March 7, 2024 in Kotaki with a technical executive from the Ministry of the Environment).

Photography 4 : Baobab leaves

Source: SEBASTIEN Garnaud, June 2006

2.3.5 Properties of bark

Bark is used to treat fever, malaria and diarrhea, and is even an anti-inflammatory, tonic for young children, for lumbago, toothache, burns, treating superficial wounds, softening the skin and making rope.

Photo 5: Baobab bark in the village of sangha, Mali

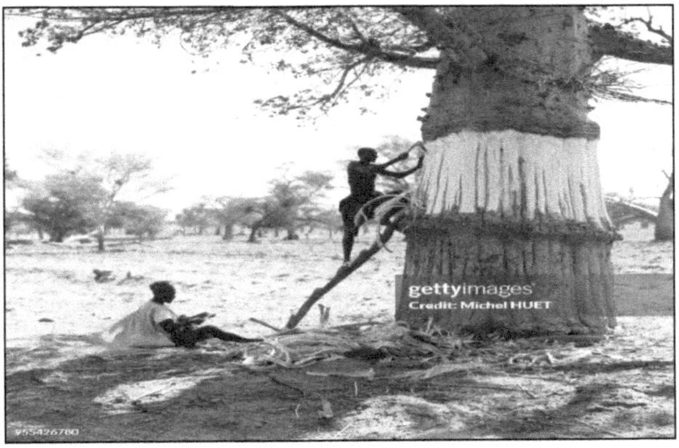

Source : MICHEL Huet

This photograph shows two men cutting the trunk of a baobab tree in the village of Sangha, Mali. These extracts are used to make multipurpose ropes for the local population, both in time and space.

"We've been using the bark of this tree for a long time to treat wounds, burns and diarrhoea. That's what we used to do here with the bark of this tree, but there are enough uses for it". (Extract from an interview in Banigoungou, March 7, 2024).

2.3.6 The properties of roots

The roots are also used as a remedy for malaria, epilepsy and agalactia (generally in combination with other plants). The same roots are used to season sauces, particularly sorrel. In fact, its roots help to break down the acidity and sourness of sorrel in place of natron.

Photography 6 : Baobab roots

Source: INECOBA, 2015

2.3.7 Flower properties

They help with childbirth and coughs, and are also used to treat anemia. Baobab flowers are used to prepare the "potash" used in human nutrition. The nectar is also sucked by children as a dessert or snack.

Baobab flowers are white, large (10-20 cm long) and hang from a stalk up to 1 m long. They are solitary or in pairs, hermaphroditic, with numerous white stamens and an ovary with 5 to 10 cells. Flowering takes place before the rainy season. Pollination is mainly carried out by bats.

Photography 7 : baobab flower

Source: E. ASSOGBADJO Achille, 2013

"I keep it in my home because it has so many uses. I use its leaves to make sauce, and I sell the pulp and seeds. In fact, these flowers are used to treat coughs, anemia and to facilitate childbirth in rural areas". (Extract from an interview on March 7, 2024 in Kotaki).

Photography 8 : baobab flower

Source: SEBASTIEN Garnaud, June 2006

Table 1: Nutritional and calorific value of baobab (in green)

Scientific name	**Part**	**Key nutrients**
Adansonia digitata	Pulp	Vit C (among the 10 richest natural products in the world)
	Seed	18% protein, 12% fat
	Leaf	11% protein, rich in Ca and Fe (> *M. oleifera*)
Balanites aegyptiaca	Oil	70% unsaturated fatty acids
Boscia salicifolia	Leaf	Rich in Iodine (> spinach)
Boscia senegalensis	Seed	20% protein, 60% carbohydrates, Zn, Fe and Ca

Maerua crassifolia	Leaf	15 to 39% protein, 7/8 essential amino acids> WHO standards, rich in Ca, Fe, Na and Ka
Sesbania leptocarpa	Leaf	Iodine, 36% protein, 8/8 essential amino acids > WHO standards
Ziziphus mauritiana	Pulp	10% protein, 80% carbohydrates, rich in Vit C
Ziziphus spina-christi	Pulp	80% carbohydrates
Sclerocarya birrea	Pulp	Vit C (4x the concentration of citrus fruits)
Crateva adansonii	Leaf	24% protein, 8/8 essential amino acids > WHO standards

Source: JOSEF Garvi, September 2021

On this point (Table n°1), we show some of the nutritional and calorific contributions of baobab according to the work of G. Josef (2021, p.18). It should be noted that Garvi's work focused on the intake of a few parts of the baobab species, such as pulp, seeds and leaves (*see Table n°1*). Given that the baobab population possesses several valuable assets for human health, and is even a source of food and income in the rural and urban world.

Table 2: Some local names for the baobab in Africa

Country	Ethnic groups	Local names
Benin	Fon	Kpassatin
	Nago	Osché
	Bariba	Sônbu
	Dendi	Kôô

	Ditamari	Sônbu, Moutomu
	Lokpa	Telou
Mali	Dogon	Oro
	Bambara	Sira
	Peulh	Babbe, boki, olohi
	Mandinke	Sira, sito
Niger	Zarma	Ko
	Haoussa	Kuka
	Peulh	Babbe, boki
	Arabic	Bu hibab
Burkina Faso	Bambara	Sira
	Jula	Bokki or yiri
	Fulfuldé and mooré	Toega

Source: field survey, March 2024

This table shows the different local spellings of baobab for a number of ethnological groups in West African countries such as Benin, Niger, Burkina Faso and Mali.

2.4. Nutritional aspects of plants

In much of the Sahel, and particularly in Mali (Bambara and Soninke communities), Burkina Faso (Mossi) and Niger (Djerma), baobab and moringa leaves are eaten daily as sauces with cereal-based meals.

Annual consumption of dried baobab leaves is estimated at 100 kg per person per year. The leaves are harvested during 3-4 months of the rainy season, dried and stored for use during 8-9 months of the dry season. (M. Larwanou, 2010, p.9).
In addition, baobab products are sold mainly in markets. The most common are leaves (fresh and dried), fruit, handicrafts and bark products (fibers).
Thus, on the basis of some work that has taken place in the Sahelian countries, certain identified species are considered to be priority and promising, including among others:

- *Adansonia digitata(*baobab*)*,
- *Vitellaria paradoxa* (shea),
- *Tamarindus indica* (tamarind),
- *Oleifera moringa* (Moringa) and
- *Ziziphus mauritiana* (jujube).

By way of example, baobab and moringa are grown in seeded, fertilized and irrigated plots for intensive, year-round production of fresh leaves. According to M. Larwanou (2010, p.8), work on intensive baobab plant production began in 1995 by ICRAF and INRAN in Niger. It used participatory research and development approaches with socio-cultural and economic studies to promote the intensification of cultivation of a number of plants.
All these tree species are rich in nutritional value and help to boost household purchasing power through the income they generate in our study area.
Leaves, fruits and by-products (pulp, seeds, kernels and fibers), bark and roots are harvested for everyday household use in food, medicine, logistics and cosmetics, as well as for sale in local markets.
The many locally-used baobab products and their unique properties offer significant commercial opportunities, notably in the food and beverage, botanical remedies and nutraceuticals, and natural cosmetics sectors.

In view of the climatic challenges ahead, the baobab production model calls on growers and even decision-makers (the State, technical and financial partners, projects, etc.) to rethink the approach that has not yet been developed.

In truth, the difficulties experienced by producers, especially in adding value to their products, call for new approaches that are much better adapted to local realities.

In fact, climate change is having a negative impact on the development of the species in question, on the one hand, and anthropogenic action in the form of deforestation is hampering it, on the other.

2.5. The state of baobab degradation in Kotaki and Banigoungou

From the graphs below, it's clear that the precious species is under enormous threat in Kotaki and Banigoungou. The various threats facing the baobab are abusive cutting of branches and bark, lack of regeneration and effective seed collection.

Graph 01 illustrates the state of degradation of the baobab at Banigoungou, showing that the level of degradation has reached 73% for very degraded, 18% for slightly degraded and 9% for degraded.

In Kotaki, too, the baobab is following the same pattern of extinction, with 75% in a very degraded state, 5% in a slightly degraded state and 20% of respondents stating that its condition was degraded following their observation (see graph 2).

Figure 1 and 2: state of baobab degradation

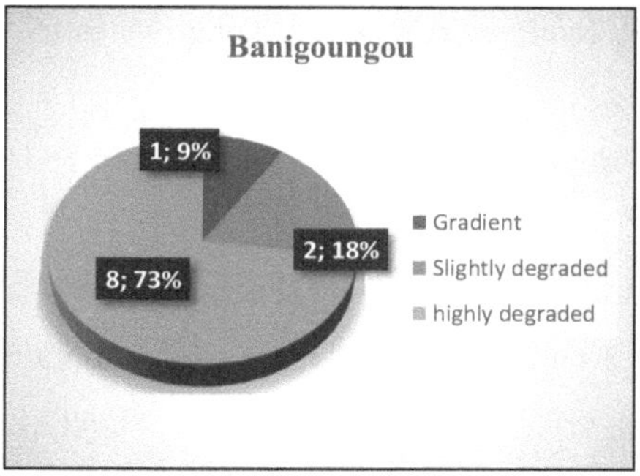

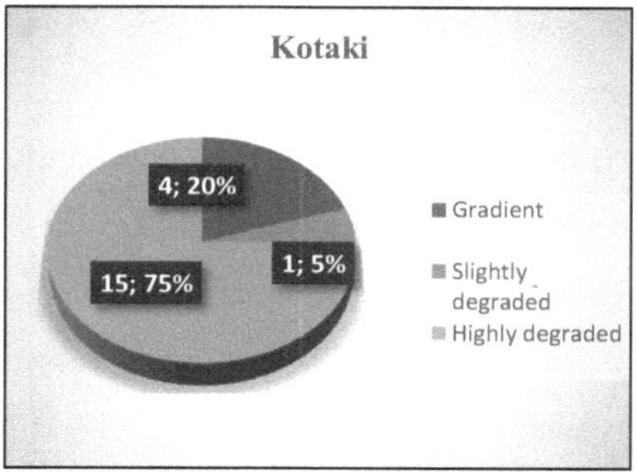

Source: field survey, March 2024

Figure 3 below shows the number of baobabs in the villages of Kotaki and Banigoungou from 20 years ago to the present day. It shows that 20 years ago, there were more than 25 plants in the Kotaki area. In the study area, the baobab is not cultivated to any great extent. It is exploited as a natural forest resource. However, uncontrolled exploitation of the plant has led to its disappearance in the rural commune of Falmey. Today, however, there are less than 10 plants in Falmey and around 20 in Banigoungou, and today there are fewer than 05 plants.

Figure 3: Number of baobabs

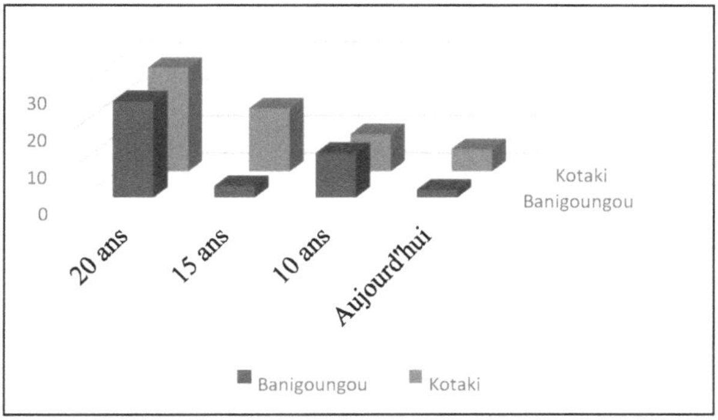

Source: field survey, March 2024

Figure 3 also shows that in the villages surveyed, there has been a clear decline, or even disappearance, of this valuable food, nutritional and pharmaceutical species.

So baobab populations can continue to provide ecological, economic and food benefits, if managed sustainably.

Baobab populations need to be managed to ensure that their genetic diversity is maintained over the long term, resources endure and baobab products remain available (DHILLION and GUSTAD 2004; SCHUMANN et al. 2010; VENTER and WITKOWSKI 2011) cited by D. AGUNDEZ (2018, p.5).

The baobab plant is important not only for its physical conservation and genetic resources, but also for its socio-economic and pharmaceutical values.

We can therefore indisputably confirm that the village, which etymologically means 04 baobabs (*Ko'itaki =Kotaki*[1]), would only be a name if the inhabitants don't get down to regenerating it.

2.6. Stakeholders involved in baobab regeneration in Kotaki and Banigoungou

A number of players have been identified in the context of the regeneration of the baobab (*Adansonia digitata L.)* population, which is indeed endangered in the rural commune of Falmey. Graph 4 shows that 11 landowners surveyed in Kotaki replied that they did not know whether any stakeholders were involved in the regeneration of the species. Then, in Kotaki (02) respondents said there were no players, and (03) in the other village, (09) surveys showed the existence of players in Kotaki and (06) in Banigoungou.

---

[1] The village of Kotaki etymologically means 04 baobabs (***Ko'itaki =Kotaki***), extract from an interview with the inhabitants of the village.

Figure 4: Evidence of the existence of the players involved

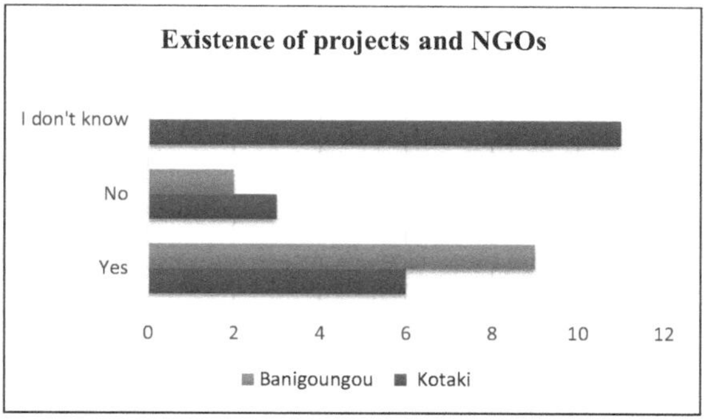

Source: field survey, March 2024

According to our data collected in the field, the populations of our two target villages affirm that these actors below have provided significant assistance to enhance the value of the baobab. Here are a few examples.

These include the NGO Aménagement des Terroirs Productions Forestières (NGO-ATPF), SWISSAID, Régénération Naturelle Assistée (RNA), Programme d'Appui à l'Agriculture Sensible aux Risques Climatiques (PARSEC), ONG-TREE AID, to name but a few.

2.6.1 Aménagement des Terroirs Productions Forestières (NGO-ATPF)

The NGO ATPF has made a major contribution to the regeneration of the baobab and oleifera moringa plants. The NGO has enabled the population of the rural commune of Falmey in general, and Kotaki and Banigoungou in particular, to benefit enormously from the baobab.

2.6.2 SWISSAID

SWISSAID is an international non-governmental organization working in Niger to improve agricultural practices (irrigated and rain-fed farming), agroecology, livestock breeding, etc. This NGO has provided enormous support for the management and regeneration of useful plants, including the baobab tree, in the rural commune of Falmey. This NGO has provided enormous support for the management and regeneration of useful plants, including the baobab, in the rural commune of Falmey.

2.6.3 Support program for climate-sensitive agriculture (PARSEC)

This support program is in line with the production and conservation of multi-purpose plants in the urban and rural world. In fact, this support program has made a number of efforts in areas such as agroecology, irrigated and rain-fed agriculture in the rural commune of Falmey.

2.6.4 Assisted Natural Regeneration (ANR)

This policy has made it possible to recover a number of endangered plants and even areas that have suffered from harmful anthropic actions such as abusive felling of plants that necessarily need to be protected, such as the baobab species. The problem of the disappearance of trees or loss of tree cover in the Sahel was addressed through the promotion of assisted natural regeneration.

So, for Larwanou and Saadou (2006, pp.14-15), assisted natural regeneration is an agroforestry method in which the natural regrowth of stumps after cutting is managed by farmers. Assisted natural regeneration is based on the existing stump in the field to be revegetated.

Assisted natural regeneration offers many benefits, including the regeneration of many species such as *Balanites aegyptiaca*, *Boscia senegalensis*, *Ziziphus spp*, *Annona senegalensis*, *Sclerocarya birrea* and *Tamarindus indica*. They provide fruit and leaves for human consumption. Some of these species are used for famine or lean periods, while others are consumed all year round.

In many cases, the fruits and leaves of certain species were the only food sources that spared communities from famine and generated income for vulnerable groups.

2.6.5 ONG-TREE AID

It promises to restore and protect the various ecosystems (ecosystem rehabilitation project in the said Park W) in Niger's Park W and to fight poverty. The aim is to promote forest governance, efficient management of natural resources and ensure food and nutritional security.) To this end, it has provided a great deal of support for the planting or regeneration of the precious baobab plant, enabling the inhabitants of Kotaki and Banigoungou to derive considerable benefit from baobab extracts.

She also worked to raise awareness among the beneficiaries of the aid to safeguard the baobab species on their territory.

As there are not so many players in the various villages, some villagers try to have it preserved in their homes and fields, as the baobab is very useful in many areas.

2.7. Contribution of projects, NGOs and initiatives

Following this work on the extinction of the baobab, field data revealed several findings on the support provided by projects, NGOs, initiatives and associations to regenerate the baobab population in the rural commune of Falmey, Dosso region.

As a result, these two villages once had huge baobab units, but these have now disappeared as a result of numerous devastating anthropic actions and other natural phenomena, as shown in graph 3.

However, to mitigate its disappearance in the rural commune of Falmey, a range of tangible and intangible contributions have been made (see figure 5).

Figure 5: Stakeholder support for plant conservation

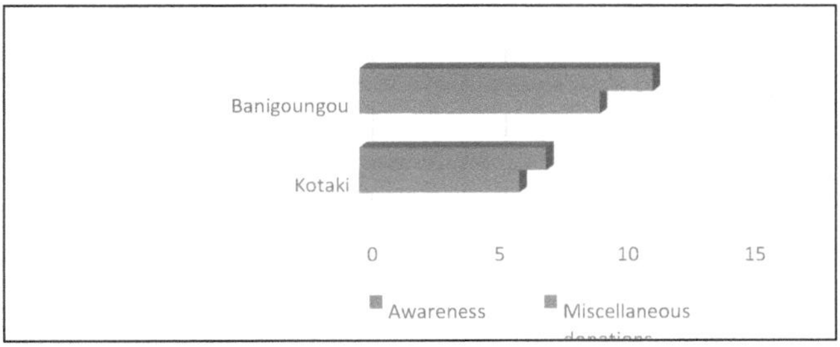

Source: field survey, March 2024

In the same vein, in order to recover its fair value, a number of players have provided support and continue to intervene by raising awareness of the baobab's food, pharmaceutical, nutritional and logistical value. They donate seeds and products to protect the baobab against the insects that hinder its development. Numerous seedlings have been made available to local residents, and planting contracts have been offered to the same beneficiaries, enabling them to set up sustainable participatory projects in their own fields, gardens and homes.

2.8. **Negative effects of the baobab's disappearance.**

The extinction of the baobab population has had a huge impact on the inhabitants of Kotaki and Banigoungou. Extracts from the baobab provide income for rural households. Given the income generated by baobab extracts, it's clear that its disappearance would have far-reaching consequences. This tree is very important in the fields of health, food, logistics and so on. Its leaves, pulp and roots are used for many purposes in both rural and urban areas. Figure 6 illustrates the effects of the baobab's disappearance according to its owners.

Figure 6: negative effects of baobab extinction

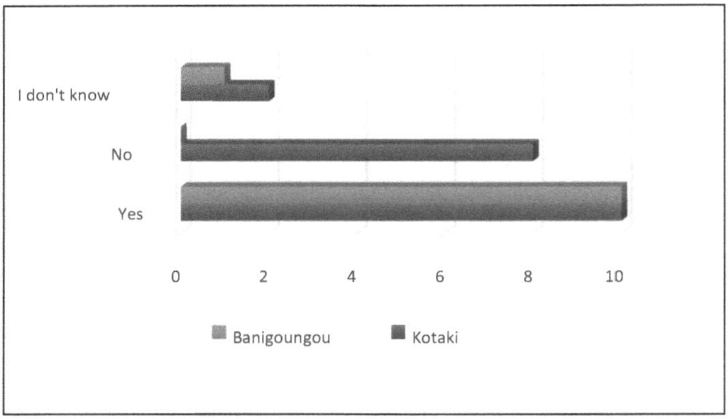

Source: field survey, March 2024

Graph 6 shows that most of the owners surveyed stated that the disappearance of the baobab has serious consequences in both rural and urban areas. In

Banigoungou (09) people and also in Kotaki confirmed that its disappearance has negative effects. But in Kotaki (07) respondents denied the effects of its extinction on their territory.

3. Discussions

The results of the field surveys showed that the baobab population appears to be under serious threat as a result of harmful human actions. In fact, in Kotaki and Banigoungou, local people report that the baobab's presence has greatly diminished. This shows that the decline is attributed to climate change and various human pressure factors, such as clearing, fires, livestock grazing, the effective collection of baobab leaves and fruit, as well as pests and diseases according to the results obtained by (AGUNDEZ et *al.* (2016) cited by D. AGUNDEZ et *al*, (2018, p.5).

Indeed, through field data in the experimental areas, this study proves the urgent need for action to protect the environment in general.

Empirical data show that the baobab population is under threat. This threat is the direct and/or indirect result of human activity. Whether it's a question of deforestation or repeated droughts, man is at the heart of it all.

According to field data, the disappearance of certain tree species is contributing to the acceleration of climate change. As a result, the respondents, whose ages ranged from 49 to over, declared that in their villages there is so much abusive cutting of roots, effective collection of leaves and seeds.

So these activities are tangible proof of its extinction in time and space. It's an endangered species that deserves to be conserved and protected in order to benefit from its various assets. In Kotaki and Banigoungou, each of the owners is starting to protect the baobab stumps that remain in their possession today.

However, which is very critical even in conservation areas such as the Parc Transfrontalier du W (Benin, Niger and Burkina Faso), there is also a decline in its presence according to studies carried out by SCHUMANN et *al.* (2012) cited by D. AGUNDEZ (2018, p.5), with a very low density and a predominance of old trees of large diameter and rare regeneration demonstrated DOUMA et al. (2010) cited by D. AGUNDEZ (2018, p.5).

Weather conditions and human pressure have caused major degradation of forest systems, but Niger is a vast territory with a diversity of agroforestry systems. This can be safeguarded in the long term, with regeneration in protected areas and a ban on intensive cutting of roots and branches. It is also essential to publicize the intervals at which baobab leaves and seeds are harvested.

Conclusion

At the end of this work, the baobab as a plant species is subject to extinction if no appropriate measures are taken in the localities studied. The same observation is made by local residents in protected areas such as the W Transfrontier Park. Hence the need for local communities, the state and decision-makers to take action to protect the species in question.

Indeed, protecting the baobab can not only help preserve the ecosystem balance, but also curb food insecurity. For example, the use of baobab products by local communities in their food systems is an important lever against certain forms of vulnerability in the Sahel in general and Niger in particular.

For example, in northern Mali, the rural population has begun to resort to importing baobab leaves to supplement its diet according to LEACH et *al.* (2011) cited by D. AGUNDEZ (2018, p.5).

This massive tree produces fruit and leaves that are traditionally eaten by local populations, but are now very rare.

Bibliography

1. ABDOU Rabiou, ABASSE Toujani, MAYAKI Allassan, SOULEY Maman Sadi, ALBERT Nikiema, JAN Svejgaard Jensen and ANDERS Raebild, 2010, *Rapport de mise en place d'un essai international de provenance d'Adansonia digitata L. (Baobab) à l'ICRISAT-Sadoré, Niger,* Numéro de titre de série Forest & Landscape Working Paper no. 48-2010, Le rapport est publié sur www.sl.life.ku.dk, ISBN 978-87-7903-495-2, 15 p.
2. Adanson Michel (1771). Description of a new tree genus called Baobab, observed in Senegal. *Hist. Acad. Roy. Sci. (Paris)* 1791: 218-243.
3. Assogbadjo A.E., Kyndt T., Chadare F.J., Sinsin B., Gheysen G., Eyog- Matig O. & Van Damme P. (2009). Genetic fingerprinting using AFLP cannot distinguish traditionally classified baobab morphotypes. *Agroforestry Systems, Agroforestry Systems* 75:157- 165.
4. Assogbadjo A.E., Glèlè Kakaï R., Chadare F.J., Thomson L., Kyndt T., Sinsin B. & Van Damme P. (2008). Folk classification, perception and preferences of baobab products in West Africa: consequences for species conservation and improvement. *Economic Botany*, 62 (1): 74-84.
5. Assogbadjo A.E. (2006). Socio-economic importance and study of the ecological, morphological, genetic and biochemical variability of baobab (*Adansonia digitata* L.) in Benin. PhD thesis. Faculty of Bioscience Engineering, Ghent University, Belgium. 213 p.
6. Codjia, J.T.C., Fonton-Kiki B., Assogbadjo A.E. & Ekué M.R.M. (2001). Le baobab (*Adansonia digitata*), une espèce à usage multiple au Bénin. Coco Multimédia, Cotonou, Benin.
7. BABOU ANDRE Bationo, NIEYIDOUBA Lamien, NICOLE Demers, SERIGNE Kandji (2009) *"Cultivation of the baobab Adansonia digitata L. (Bombacaceae) in market garden beds: a method to simplify harvesting and promote propagation in the Sahel"*, 9 p.

8. Convention on International Trade in Endangered Species of Wild Fauna and Flora : Seventeenth Session of the Conference of the Parties Johannesburg (South Africa), September 24 - October 5, 2016, CoP17 Prop. 58. 8 p.
9. DIA Abdoulaye, DUPONNOIS Robin, 2012, *"the Great Green Wall"*. DOI : 10.4000/books.irdeditions.3343, 23 p.
10. GABAR Diop Aïda, SAKHO Mama, DORNIER Manuel, CISSE Mady, REYNES Max (2005) *"Le baobab africain (Adansonia digitata L.): principales caractéristiques et utilisations"*, 15 p.
11. Intergovernmental Panel on Climate Change (IPCC), 2007, Climate Change 2007. Contribution of Working Groups I, II and III to the Fourth Assessment Report of the Intergovernmental Panel on Climate Change, Geneva, IPCC, 103p.
12. ISIDORE Kouwonou, BBC Afrique, February 19, 2024, *"Baobab leaves are a nutritional treasure"* available on the Internet [online]: www.bbc.com accessed August 28, 2024 at 10:33 pm.
13. JOSEF Garvi, 2021, Sahara Sahel Foods, Parliamentary Seminar on September 28, 2021, *"Nutrition through the indigenous food trees of Niger"* in Niger, Niamey, 32 p.
14. Kerharo J., Adam J.G., La pharmacopée sénégalaise traditionnelle - Plantes médicinales et toxiques, Vigot Frères, Paris, France,1974.
15. LARWANOU Mahamane, 2010, *" Espèces prioritaires et options agroforestières pour la mise en valeur de la Grande Muraille Verte "* In : Le projet majeur africain de la Grande Muraille Verte : Concepts et mise en œuvre [online]. Marseille: IRD Éditions, 2010 (generated on October 18, 2023). Available on Internet: <http://books.openedition.org/irdeditions/2113>. ISBN : 9782709918244. DOI : https://doi.org/10.4000/books.irdeditions.2113.
16. MASSAOUDOU Moussa, LARWANOU Mahamane and MAHAMANE Saadou, 2015, *"Caractérisation des peuplements ligneux des parcs à Faidherbia albida (Del) A. Chev. et à Prosopis africana (Guill.,Perrot et*

Rich.),Taub,Du Centre-Sud Nigérien", in Journal of Applied Biosciences, December 2015 DOI: 10.4314/jab.v94i1.6, 18 p.

17.SEHENO Andriantsaralaza, MIGUEL Pedrono, JACQUES Tassin, ROGER Edmond, PASCAL Danthu,(2012, "Baobabs de Madagascar:un anachronisme de la dispersion?" 10 p.

18.Office National d'Edition et de Presse (ONEP), 2023, "*Promotion de baobab dans le canton de Konni : une initiative à implémenter dans les autres zones du pays",* available on the Internet [online]: www.lesahel.org accessed on 28 August 2024 at 22h 10.

19.RANKOVIC, A. C. PACTEAU and L. ABBADIE, 2012, "*Ecosystem services and interscalar urban adaptation to climate change: an attempt at articulation".* Hors-série 12, VertigO - la revue électronique en sciences de l'environnement [Online] URL: http://vertigo.revues.org/11851 DOI: 10.4000/vertigo.11851, accessed July 26, 2024. DOI : 10.4000/vertigo.11851.

20.SEBASTIEN Garnaud, 2006, international conference GRECOREV: Gestion Concertée des Ressources naturelles et de l'Environnement-du local au mondial: pour un dialogue entre les chercheurs, sociétés civiles et décideur. Université de Versailles Saint-Quentin-en-Yvelines (June 26-28, 2006) "*Programme d'un projet d'Appui et d'aide à la sauvegarde de la forêt de baobabs de Nuégkokh (Sénégal) Association Chêne et Baobab",* futura-sciences.com/planete.

21.SEBASTIEN Garnaud, 2006, Elaboration of the Baobécole project, an environmental education and scientific awareness program based around the baobab tree in Africa.

22.SERGE Paugam (dir.), L'enquête sociologique, PUF, coll. " Quadrige Manuels ", 2010, 458 p., EAN : 9782130555032, Référence électronique Elieth P. Eyebiyi, " Serge Paugam, L'enquête sociologique ", Lectures [En ligne], Les comptes rendus, online June 24, 2010, accessed September 09, 2024. URL :

http://journals.openedition.org/lectures/1055 ; DOI : https://doi.org/10.4000/lectures.1055.

23. THEAU, J.P., P. Cruz, D. Fallour, C. Jouany, E. Lecloux and M. Duru, 2010, Une méthode simplifiée de relevé botanique pour une caractérisation agronomique des prairies permanentes, Fourrages, 201, pp. 19-25.
24. BARDOT J. (2010), "Mener un entretien en face à face", in Paugam S. (dir.), L'enquête sociologique, Paris, PUE p. 115-141.
25. BEAUD S. AND WEBER F. (2010). Guide de l'enquête de terrain: produire et analyser des données ethnographiques, 4eéd. Paris, La Découverte.
26. BECKER H. S. (2002). Les ficelles du métier: comment conduire sa recherche en sciences sociales, Paris, La Découverte.

BERTAUX D. (2005). L'enquête et ses méthodes: le récit de vie, Paris, Armand Colin.

27. BERTHIER N. (2010). Les techniques d'enquête en sciences sociales: méthodes et exercices corrigés, 4eéd. Paris, Armand Colin.
28. BLANCHET A. AND GOTMAN A. (2007). L'enquête et ses méthodes: L'entretien, 128eéd. Paris, Armand Colin.
29. ASSOGBADJO E. Achille, the African baobab (*Adansonia digitata* L.): description, use and potential for development, 68.p

APPENDICES 1

Questionnaire

Usefulness of endangered plant species in rural Niger: the case of the baobab "*Adansonia digitata*" in Kotaki and Banigoungou.

---

1. Your first and last name: ....
2. How many types of Adansonia digitata species do you know in this area?

   ....

3. Why did you plant it?

   ....

4. What's its purpose?

   ....

5. How far has it spread in the area?
   a. Gradient...
   b. Slightly degraded...................................
   c. Very degraded..................................
6. In this area, how many A. digitata plants do you have in your field or locality?
   a) 20 years back in time......................................
   b) 15 years back in time......................................
   c) 10 years back in time...................................
   d) Today......................................
7. What are the main threats to the Adansonia digitata population?
   a) The collection and massive exploitation of its leaves.................................
   b) Fruit picking...
   c) Seed collection...
   d) Final felling of trees for bark collection..............

e) Water deficit...........................

f) Searching for arable or pasture land...............

g) If other, please specify..

8. Are there any players (projects/NGOs/initiatives/associations) involved in the regeneration of this species?...................................

9. If so, which ones (names)?

....

10. What support do these players provide for Adansonia digitata regeneration?

a) Donate seeds................................

b) Awareness ....

c) If other, please specify...

11. Does the extinction of this species have a negative impact on the lives of people living in the study area?

....

12. If so, who, what and what do you think are the negative impacts?

....

13. What solutions do you propose for the regeneration of this important species?

....

A ....*le* ....

APPENDIX 2

Interview guide

Usefulness of endangered plant species in rural areas of Niger: the case of the baobab "*Adansonia digitata*" in Kotaki and Banigoungou

---

1. Name of project or NGO:... ....
2. How long will you be in the area?
3. Why are you fighting the extinction of this species?

....

4. How can Adansonia digitata help you?

....

5. What are the main threats to the Adansonia digitata population?

....

6. What support have you provided for the regeneration of Adansonia digitata?

.... Does the extinction of this species have a negative impact on the lives of people living in the study area?

....

7. If so, what do you consider to be the negative impacts?

....

8. What solutions do you propose for the regeneration of this important species?

Date ....

Kolibri 360 GmbH
Ferdinand-Jühlke-Straße 7
99095 Erfurt, Deutschland
produktsicherheit@kolibri360.de

Druck:
CPI Druckdienstleistungen GmbH
im Auftrag der
Kolibri 360 GmbH
Ferdinand-Jühlke-Str. 7
99095 Erfurt